CATERPILLAR SIXTY
PHOTO ARCHIVE

CATERPILLAR SIXTY

PHOTO ARCHIVE

Photographs from the
Caterpillar Inc. Corporate Archives

Edited with introduction by
P. A. Letourneau

Iconografix
Photo Archive Series

Iconografix
P.O. Box 18433
Minneapolis, Minnesota 55418 USA

Library of Congress Card Number 93-78195

ISBN 1-882256-05-0

93 94 95 96 97 98 99 5 4 3 2 1

Cover and book design by Lou Gordon

Printed in the United States of America

PREFACE

The histories of machines and mechanical gadgets are contained in the books, journals, correspondence and personal papers stored in libraries and archives throughout the world. Written in tens of languages, covering thousands of subjects, the stories are recorded in millions of words.

Words are powerful. Yet, the impact of a single image, a photograph or an illustration, often relates more than dozens of pages of text. Fortunately, many of the libraries and archives that house the words also preserve the images.

In the Photo Archive Series, Iconografix reproduces photographs and illustrations selected from public and private collections. The images are chosen to tell a story...to capture the character of their subject. Reproduced as found, they are accompanied by the captions made available by the archive.

The Iconografix Photo Archive Series is dedicated to young and old alike, the enthusiast, the collector and anyone who, like us, is fascinated by "things" mechanical.

ACKNOWLEDGMENTS

The photographs which appear in this book were made available by the Caterpillar Inc. Corporate Archives. We are most grateful to Caterpillar Inc., and sincerely appreciate the cooperation of Joyce Luster, Corporate Archivist.

We also wish to thank Ziegler Inc., Minnesota-based Caterpillar dealer, for its loan of supplemental photographs.

INTRODUCTION

In April 1925, Holt Manufacturing Company and C.L. Best Tractor Company merged to create Caterpillar Tractor Company. Each company brought an existing line of track-laying tractors to the union. Holt, who had built its first such machine in 1904, contributed the 2 Ton, 5 Ton, and 10 Ton. Best, who had built its first such unit in 1913, contributed the Thirty and Sixty.

Best had introduced the Sixty in 1919. The company rated its performance at 60 belt and 35 drawbar horsepower. The Sixty was fitted with a vertical 4-cylinder gasoline engine, manufactured by Best, that operated at 650 rpm. It featured 6.5 x 8.5 inch bore and stroke. The tractor's 2-speed transmission offered operating speeds of 1.875 and 2.625 mph. Maximum drawbar pull was rated at 11,000 lbs. In 1921, the Sixty was fitted with a 3-speed transmission, which, along with modifications to the engine's carburetion and ignition systems, resulted in improved performance. By 1924, the Sixty developed a maximum 72.51 brake horsepower, and 12,360 lbs drawbar pull.

The Sixty's greatest strength was its ability to operate effectively in almost any soil condition. It offered greater traction and flotation than were available with wheel tractors. Consequently, the track-equipped Sixty was widely adopted for agricultural, logging, mining, earthmoving, road building, and other construction applications.

The Sixty changed little after the formation of Caterpillar, and remained in production until 1931. It was manufactured both at San Leandro, California (A Series) and Peoria, Illinois (PA Series). In total, and including Best production, 5,413 units were built at San Leandro; 13,516 units were built at Peoria. Enthusiasts should note that the Caterpillar Inc. Corporate Archives include hundreds of photographs of the Sixty, but very few from the years before the Holt and Best merger. For this reason, the Best Sixty is not included as a part of this book.

In 1931, after years of research and an investment of $1 million, Caterpillar introduced its first diesel engine, a 4-cylinder design with 6.125 x 9.25 inch bore and stroke. The earliest production diesel tractors were constructed on the Sixty chassis, and were designated the Diesel Sixty. The performance characteristics of the Diesel Sixty were decidedly impressive: 77.08 maximum brake horsepower; 65.11 maximum drawbar horsepower; 11,991 lbs drawbar pull; and a record setting horsepower to fuel efficiency ratio of 13.87 horsepower hours per gallon of fuel. In spite of the economic down turn of the Depression years, the new tractor was an immediate success. Its performance and efficiency outshone that of any competitor. We are very pleased to include photographs of the historic Diesel Sixty, the first American built diesel tractor.

Caterpillar Sixty three-quarter front view. November 1930.

Sixty Logging Cruiser. June 1928.

Sixty fitted with steel bumper, shield, and winch. December 1926.

Sixty fitted with factory cab, lights, and Willamette winch.

18

Sixty fitted with factory cab, lights, and Willamette winch.

Diesel Sixty. September 1931.

Diesel Sixty.

Sixty hauling 17,000 feet of mahogany. Dunkwa, West Africa.

Proud loggers pose for the camera. Eureka, California. October 1928.

Sixty with LaPlant-Choate bulldozer building a logging road. September 1930.

Sixty ripping up the ground ahead of a grader. Mt. Hoos National Forest. 1930.

Sixty with Athey Fairlead wheelers. Westwood, California. October 1928.

Sixty with LaPlant-Choate bulldozer building a logging road for Crossett Western Timber Co. March 1930.

Sixty skidding logs up a small mountain stream turned into a logging trail. Kelso, Washington. July 1930.

Sixty logging with a pan near Gilmer, Washington. May 1931.

Sixty chunking out the right-of-way for a new logging road. Astoria, Oregon. 1929.

Sixty logging with Fairlead wheels. Kinzua, Oregon. 1930.

Sixty with Athey wheels, during time studies carried out by the U.S. Forestry Service. 1930.

Sixty hauling logs with high wheeler. 1926.

Sixty skidding logs in Sequoia National Park, California. September 1930.

Sixty hauling 1,250 ties from a lumber operation in Hudson, Ontario. April 1929.

Sixty chunking out the right-of-way for new logging road. Astoria, Oregon. 1929.

Wide Track Sixty pulling a disk harrow in land reclaimed from Holland's Zuider Zee. 1931.

Sixty with Dinuba vine cutter. California. January 1927.

Sixty furrow plowing in Hawaiian sugar cane field.

Sixty cross-scarifying land to be planted in Persian lime trees. Homestead, Florida. August 1935.

Sixty clearing mesquite bushes with specially built brush rake. Pima Indian Reservation, Arizona. June 1930.

Caterpillar Diesel Sixty (Serial No. 1-C-2) rips earth 7 to 9 inches deep with a Brenneis chisel. Woodland, California. April 1932.

Sixty with subsoiler tearing out old peach orchard. California. December 1929.

Sixty fitted with multiple hitch. Pullman, Washington. September 1931.

Sixty and 8-bottom John Deere plow working in the Palouse Hills. Pullman, Washington. October 1931.

Sixty plowing in muckland. Hollandale, Minnesota. September 1926.

46

Sixty with 36 foot duck-foot cultivator. Great Falls, Montana. May 1928.

Sixty plowing under hemp. July 1930.

Sixty pulling 4 ton "super-plow." The unit cuts a furrow from 36 to 42 inches deep and 34 inches wide. Huntington Beach, California. January 1932.

Picking up and threshing rice bundles with Sixty and Holt Model 36 Western Harvester. C.L. Best Ranch. Colusa, California. October 1927.

Sixty with Rumley separator converted into a pickup harvester by addition of two special feeders and Holt Model 75 motor mounted on Caterpillar tracks. Colusa, California. October 1927.

Sixty pulling an Advance-Rumley combine on 60% grade. Washington.

Sixty with Model 30 Holt harvester. Pullman, Washington. August 1928.

Sixty fitted with asparagus chopper. California. January 1927.

Rear view of asparagus chopper. California. January 1927.

Sixty operating a hay chopper. November 1930.

Sixty pulling four grain binders. Alicia, Michigan. September 1927.

Sixty running stationary harvester. Onawa, Iowa. September 1929.

Sixty planing the bumps out of an old road with a scarifier attached to a grader. Alameda County, California. July 1928.

Sixty with McMillan dirt mover. Catalina Island, California. May 1930.

Sixty and LaPlant-Choate bulldozer stripping down a bank. March 1932.

Sixty leveling for a new stretch of pavement along the Golden State Highway. Fresno, California. August 1930.

Caterpillar Sixty Diesel (Serial No. 1-C-7) pulling a blade grader. Georgia State Highway Department. May 1937.

One Caterpillar Sixty pulls an elevating grader, while the other pulls a wagon.

Three Sixtys hauling crawler wagons from a Caterpillar drawn elevating grader. Federal Highway 16, east of Sheridan, Wyoming. September 1931.

Sixty equipped with Master backfiller operated by a Willamette-Ersted double drum winch. September 1930.

Sixty leveling land. Chicago, Illinois.

Sixty pulling a Lakewood subgrader on Bankhead Highway. Lincoln, Alabama. September 1931.

Building roads in Tennessee with Sixty and Euclid scrapers. May 1929.

A Sixty with Russell elevating grader. A Sixty and Western crawler wagon.

Caterpillar Sixtys pulling 6 yard Streich wagons. A Sixty elevating grader and a Russell grader complete the job of grading. Hume, Illinois. July 1930.

Sixty and No. 68 Killefer revolving scraper. Arkansas. August 1928.

Sixty equipped with LaPlant-Choate bulldozer, on state highway project near Graham, North Carolina. July 1928.

Sixty Logging Cruiser with bumper pulling a 10 foot grader with back sloper. Toledo, Ohio. June 1928.

Sixty with Russell blade grader working at Hill City, South Dakota. July 1929.

Sixty leveling land with electrically operated 12 yard scraper. Stockton, California. January 1930.

Sixty and No. 42 elevating grader at work on a road job near Calvin, North Dakota. September 1937.

Sixty and Russell elevating grader.

Sixty moving dirt from a 40 foot cut. Woodbury County, Iowa. June 1928.

Leveling ground for construction of a glass factory with Sixty and Russell elevating grader. South San Francisco, California. June 1928.

Sixty with Athey 3-way dump wagon. January 1928.

Sixty building roads with blade grader. Minnehaha County, South Dakota. October 1927.

Caterpillar Sixty with six Baker Maney Model DR Scrapers. Columbia City, Indiana. August 1931.

Sixty equipped with experimental rear power take-off, and power operated Russell elevating grader. Arkansas. February 1929.

Sixty pulling a 7 yard crawler wagon. Bloomington, Minnesota. August 1928.

Sixty with Gardner loader in Country Club Manor, Long Beach, California. May 1929.

Casting-in with a Caterpillar Sixty and No. Sixty elevating grader. Brown County, South Dakota. August 1931.

Sixty with two LaPlant-Choate crawler wagons hauling from a steam shovel. Minnesota. September 1930.

Sixty and disk-brush on ditch banks. July 1930.

Caterpillar Sixtys photographed at East Peoria plant following 1927 flood damage.

Pushing trash into a swamp with Sixty equipped with bulldozer. Newark, New Jersey. October, 1931.

Sixty hauling chariot-type wagon on the Hoover Dam project. May 1931.

Sixty with McMillan scraper sloping steep banks. July 1928.

Sixty acts as a switch engine for Great Northern Railroad. August 1931.

Sixty fitted with homemade side boom winch lowering 22 inch gas line into ditch. Pittsburgh, California. March 1930.

Sixty hauling one mile of snow fence. Minnesota. April 1929.

Sixty with LaPlant-Choate bulldozer used for construction of drainage ditches by Pennsylvania Railroad. Metuchen, New Jersey. December 1929.

Sixty bulldozing dirt over culvert. Dublin Canyon, California. December 1927.

Hauling gold ore with Caterpillar Sixty Snow Special and LaPlant-Choate crawler wagon. Alma, Colorado. April 1931.

Sixty equipped with Allsteel side boom lifting 24 foot section of 22 inch pipe. Between Bakersfield and Los Angeles, California. October 1930.

Caterpillar Diesel Sixty (Serial No. 1-C-4) hauling two Euclid wagons fitted with 12 inch flareboards. Antacostia, Virginia. January 1932.

Sixty with LaPlant-Choate snow plow. Itasca, Illinois. March 1931.

Clearing runway at Rochester, New York Municipal Airport with Sixty and Wausau plow. March 1931.

Sixty hauling and placing vans on an amusement lot; setting up for Elks convention. Seattle, Washington. July 1931.

Wm. H. Zeigler Co. exhibit at the 1927 Minnesota State Fair.

Sixty hauling sugar cane to the mill. Puerto Rico. 1930.

Sixty operating a hot mix plant on Highway 30. Rawlins, Wyoming. September 1931.

Caterpillar tractors working on levee construction. March 1930.

Sixty leveling sand hills along Pacific Ocean. California. December 1929.

Hauling molding sand with Sixty and dump wagon. East of Evansville, Indiana. July 1929.

Sixty unloading a locomotive from a ship. Ford Rubber Plantation. Boa Vista, Brazil. 1931.

Laying natural gas line with Sixty equipped with side boom.

Caterpillar Sixty Diesel (Serial No. 1-C-2) pumping water for irrigation near Woodland, California. August 1940.

Digging slush pits for rotary well with Sixty and Ball Fresno wagon. May 1930.

Sixty used to move oil field equipment. Centralia, Illinois. February 1938.

Sixty with Hopper hoist in oil fields near Bakersfield, California.

Sixty furnishes all the power for transporting, setting up and operating a Star Oil rig. Saginaw, Michigan. October 1928.

Sixty pulling an Ateco dirtmover and sheepfoot tamper, at work on West Seattle Reservoir. White Center, Washington. June 1931.

Sixty with electrically operated bulldozer moving sand dunes. Sunset District, San Francisco. May 1927.

Three Caterpillar Sixtys and one Thirty laying long distance cable for Southwestern Bell. Between Kansas City and Joplin, Missouri. March 1931.

Sixty with Schramm air compressor mounted on Athey wheels. Harlingen, Texas. April 1931.

Sixty leveling land after cactus has been removed. Pima Indian Reservation, Arizona. March 1931.

Caterpillar Sixty Diesel (Serial No. 1-C-4), built October 28, 1931, leveling land for new Pan American Airways base. Baltimore, Maryland. November 1936.

Sixty with bulldozer working on the Akron, Ohio airport. September 1930.

Sixty with Willamette winch used to construct 2000 foot inclined tramway. British Columbia.

A specially built machine based on a Caterpillar Sixty, used for preparing ground for drying salt. Alviso, California. September 1927.

Caterpillar Sixty with train of wagons hauling dirt at the site of a new observatory. Guatemala City, Guatemala. December 1929.

Sixty with Baker snowplow. March 1931.

Sixty with Hauser brush cutter.

Sixty with bulldozer backfilling sewer trench. Grant Park, Chicago, Illinois. November 1928.

Caterpillar Sixty to the rescue! This load of livestock was snowed in. The Sixty saved them. Moro, Oregon. February 1933.

Sixty and McMillan hydraulic scraper grading slopes of Forest Lawn Memorial Park. Glendale, California. April 1929.

Sixty cleaning up the Oak Street Beach. Chicago, Illinois. May 1931.

Sixty grading an airfield. Peoria, Illinois.

Sixty and bulldozer used for various jobs on a Eureka, Washington ranch. May 1954.

Sixty pulling down the steeple base of the Old Brown Street Methodist Church. Knoxville. April 1927.

Moving a house with a Caterpillar Sixty.

Sixty and Russell grader working on a Minneapolis street.

Sixty hoisting a 34 ton safe deposit vault door from the basement of a St Louis, Missouri bank. June 1929.

The Iconografix Photo Archive Series includes:

JOHN DEERE MODEL D Photo Archive	*ISBN 1-882256-00-X*
JOHN DEERE MODEL B Photo Archive	*ISBN 1-882256-01-8*
FARMALL F-SERIES Photo Archive	*ISBN 1-882256-02-6*
FARMALL MODEL H Photo Archive	*ISBN 1-882256-03-4*
CATERPILLAR THIRTY Photo Archive	*ISBN 1-882256-04-2*
CATERPILLAR SIXTY Photo Archive	*ISBN 1-882256-05-0*
TWIN CITY TRACTOR Photo Archive	*ISBN 1-882256-06-9* *(Sept. 93)*
MINNEAPOLIS-MOLINE U SERIES Photo Archive	*ISBN 1-882256-07-7* *(Sept. 93)*

The Iconografix Photo Archive Series is available from direct mail
specialty book dealers and bookstores throughout the world,
or can be ordered from the publisher.
For information write to:

Iconografix
P.O. Box 18433
Minneapolis, Minnesota 55418 USA